Black Widow Spiders

BY ELIZABETH RAUM

AMICUS HIGH INTEREST   AMICUS INK

Amicus High Interest and Amicus Ink are imprints of Amicus
P.O. Box 1329, Mankato, MN 56002
www.amicuspublishing.us

Copyright © 2016 Amicus. International copyright reserved in all countries. No part of this book may be reproduced in any form without written permission from the publisher.

Raum, Elizabeth, author.
 Black widow spiders / by Elizabeth Raum.
    pages cm. – (Poisonous animals)
 Audience: K to grade 3.
 Includes bibliographical references and index.
 ISBN 978-1-60753-784-7 (library binding)
 ISBN 1-60753-784-2 (library binding)
 ISBN 978-1-60753-883-7 (ebook)
 ISBN 978-1-68152-035-3 (paperback)
 1. Black widow spider–Juvenile literature. 2. Poisonous spiders--Juvenile literature. 3. Children's questions and answers. I. Title.
 QL458.42.T54R38 2016
 595.4'4–dc23

                                   2014033268

Editor: Wendy Dieker
Series Designer: Kathleen Petelinsek
Book Designer: Heather Dreisbach
Photo Researcher: Kurtis Kinneman
Photo Credits: Scott Camazine Cover; Alamy/John Cancalosi 5; Alamy/Scott Camazine 6; Corbis/John Giustina 13; Corbis/Jeff Howe/Visuals Unlimited 22; Science Source / James H Robinson 21, 24–25; Science Source/Scott Linstead 10; Science Source/Francesco Tomasinelli 28; Shutterstock/Anton Foltin 17; Shutterstock/tea maeklong 18; Shutterstock/sritangphoto 27; Superstock/Minden Pictures 8–9; Wikipedia/Calistemon 14

HC 10 9 8 7 6 5 4 3
PB 10 9 8 7 6 5 4 3 2 1

Table of Contents

Yikes! A Black Widow!	4
Caught in a Web	11
Staying Alive	16
Baby Spiders	20
Some Good News	26
Glossary	30
Read More	31
Websites	31
Index	32

Yikes! A Black Widow!

Darkness falls. A black **widow** spider hangs upside down in her web. She waits. Her feet touch the web. They can tell if something touches it. Suddenly, the web shakes. It's a sign. The black widow attacks! She rushes forward. Is it something tasty? It's a bee! She wraps the **prey** in silk. It can't escape now!

The female black widow spider waits for a meal to get trapped in her web.

The spider eats the bee
she caught in her web.

 Can black widow **venom** kill people?

The black widow bites the bee. The spider's fangs send venom, a kind of poison, into the bee. The venom is strong. It is 15 times stronger than a rattlesnake's. A black widow's bite can kill insects, lizards, and small animals. But don't worry. Black widows do not hunt people. They only bite to defend themselves.

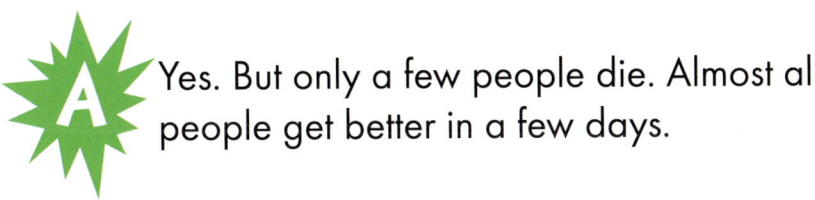 Yes. But only a few people die. Almost all people get better in a few days.

Female black widows are about 1 inch (2.5 cm) long. They are black with a bright red mark on their **abdomen**.

Males look much different. They are smaller. They also are differently colored. Males are brown or tan. Some have red or tan stripes on their backs. Others have yellow spots.

The female black widow (bottom) is larger than the male (top).

Black widows make their webs in holes and hunt at dusk.

 Does the spider stick to her web?

Caught in a Web

Black widows are weavers. They make rough and sticky webs. They spin webs in corners. They spin webs between rocks, leaves, or trash. They spin webs in old tin cans. The webs are strong. Insects and other bugs crawl onto the sticky webs.

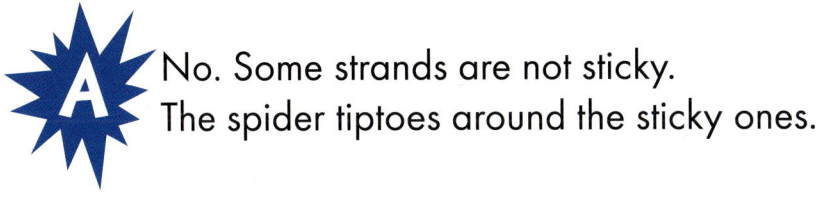
No. Some strands are not sticky. The spider tiptoes around the sticky ones.

The black widow hides during the day. At night she traps bugs like flies, moths, and mosquitoes in her web. She wraps them in silk and bites them. The venom kills the prey. It also turns the prey's insides soft and juicy. Then the black widow sucks the juices. She eats the rest.

 Do males hunt too?

A black widow waits for prey in her web.

 Yes. They also use webs and venom to kill insects. But their venom is not strong enough to harm people.

Most of the time, black widows catch insects. Sometimes they catch bigger things. A black widow might catch a snake or lizard in her web. It is too big to wrap in silk, but the spider will bite it. The animal won't be able to move. Its body will swell. Then it will die. Its insides will turn juicy. The spider will eat it.

Black widow spiders sometimes catch small lizards.

Staying Alive

Black widows live all over the world. Three **species**, or kinds, live in the United States. Their name tells where they live. The western black widows live in the west. Northern black widows live in the northern part of the country. Can you guess where southern black widows live?

The western black widows live in desert areas of the southwestern United States.

A praying mantis is one of only a few animals that eat black widows.

Most **predators** leave black widows alone. The venom is a strong defense. The female's red mark is a warning. It warns other animals to stay away. Birds often eat spiders, but they don't eat black widows. Eating black widows makes many birds and lizards sick. Only alligator lizards, praying mantises, and mud-dauber wasps eat black widows.

Baby Spiders

Black widow spiders live alone. In the spring, the males stop eating. They look for females. And then they **mate**. The female spins about 10 egg sacs. A black widow's egg sac holds about 300 eggs. The female guards them until the eggs hatch. They hatch in about 14 days.

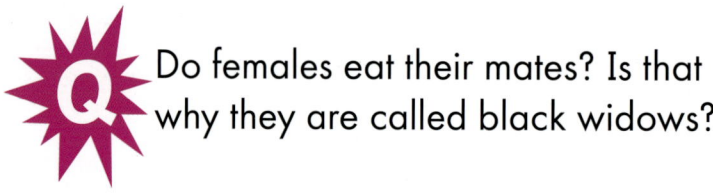
Do females eat their mates? Is that why they are called black widows?

An egg sac made of spider silk holds black widow eggs.

 That is why they are called black widows. But it is not true. Only one species in Australia actually does this.

Tiny spiderlings hatch and crawl out of the egg sac.

 Why do so many babies die?

The spiderlings are tiny. Their bite is harmless. They are tan or white. As they grow, the babies **molt**. This means that they shed their **exoskeleton**, or outside shell. They change color. At first, there are hundreds of babies. Only a few from each egg sac live for more than a month.

 They eat each other.

After three weeks, the spiderlings climb to a high place. They spin silk threads. They float away on the wind. When they land, they build their own webs.

Females are fully grown in 3 to 8 months. They live up to 18 months. Males grow up faster. They only live for 3 to 7 months.

Baby spiders spin their own webs and capture their own food.

Some Good News

Scientists study black widow webs. The silk, or thread, is very strong. It is stronger than steel. It is stretchy, too. Scientists want to learn how to make it. Someday we may use fake spider silk to make cables for bridges. It may even be used to make better bandages. Strong, stretchy silk has many uses.

Scientists want to learn to make spider silk. It could be used for bridge cables.

This black widow will not bite hikers if it is left alone.

 What if I'm bitten?

28

Black widows can be helpful. They eat pesky mosquitoes. They eat flies and other bugs. They can be harmful, too. Their bite makes people sick. Sometimes it kills. But remember that black widows don't want to hurt us. They just want to be left alone to eat bugs.

Call 911! Put ice on the bite. Doctors have medicine called **antivenin** that can help.

Glossary

abdomen The part of the body that holds the spider's guts, hearts, and silk glands.

antivenin Medicine used to treat venomous spider bites.

exoskeleton A hard outside shell that supports and protects the spider's body.

mate To come together to create children.

molt To shed the outside skin or shell so a new one grows.

predator An animal that hunts another for food.

prey An animal that is hunted for food.

species A kind or group of animals that share certain characteristics.

venom A kind of poison produced by some animals, like black widow spiders.

widow A woman whose husband has died.

Read More

Kopp, Megan. *Black Widow Spiders.* New York: AV2 by Weigl, 2012.

Markle, Sandra. *Black Widows: Deadly Biters.* Minneapolis: Lerner, 2011.

Marsico, Katie. *Black Widow Spiders.* New York: Children's Press, 2014.

Websites

Black Widow | National Geographic Kids
kids.nationalgeographic.com/animals/black-widow

Hey! A Black Widow Spider Bit Me! | KidsHealth
kidshealth.org/kid/ill_injure/bugs/black_widow.html

Spider: Black Widow | Exploring Nature Resource
www.exploringnature.org/db/detail.php?dbID=43&detID=1154

Every effort has been made to ensure that these websites are appropriate for children. However, because of the nature of the Internet, it is impossible to guarantee that these sites will remain active indefinitely or that their contents will not be altered.

Index

abdomen 8
babies 22, 23, 25
color 8, 23
egg sacs 20, 23
hunting 7, 12
life span 25
males 8, 12, 20, 25
markings 8, 19
mating 20

predators 19
prey 4, 7, 11, 12, 15
silk 4, 12, 15, 25, 26
size 8
spiderlings 23, 25
venom 6, 7, 12, 13, 19
webs 4, 10, 11, 12, 13, 15, 25, 26

About the Author

Elizabeth Raum has worked as a teacher, librarian, and writer. She enjoyed doing research and learning about poisonous animals, but she hopes never to find any of them near her house! Visit her website at: www.elizabethraum.net